Kindergarten Plant Unit Study

How to use this Study

In each Science study, there are 2-3 math pages; vocabulary, sight words, and a myriad of other activities and pages.

These Units were designed to take one week, in monthly conjunction with our State and President Studies; but you can extend the activities if you wish. I will soon have additional math lessons for purchase, and we are working on developing a Big Book of Unit Study, for K-6.

If you do follow the week approach, I recommend Math on MWF, Vocab and Sight Words Daily, and the other Activities Daily.

This is an open and go book, but planning for me always works better.

Good luck on your Homeschooling Journey!!

Note:

There are some pages that mention to cut out.

Please copy these pages if you wish to cut them out, so you still have the pages on the back.

Table of Contents

Page Number

3-4 Sight Words

5-7 Vocabulary

8 Plant Facts

9 Lines

10-12 Tracing

13-14 Parts of a Flower

15-16 Parts of a Tree

17-19 Photosynthesis

20-23 Lifecycle of a Tree

24 Matching

25-29 Math

30 5 Senses Hunt

31 Bingo

32-35 Tic Tac Toe

36-37 Tomato Sauce Recipe

PLANT SIGHT WORDS

Leaf	Flower
Tree	Sun

PLANT SIGHT WORDS

WEEK 1
For the first week, show these to your child every day. Make sure they are looking at the card, and go through all of them 3 times. Don't ask them to tell you what it is, just tell them what it is.

WEEK 2
1st round- Ask them what the words are. If they don't know, tell them.

2nd round- Put them on the ground in a circle. Ask them to jump to "this word." See if they are right!

3rd round- If they are still having trouble, explain to them why the word is the word. Have them remember based on sounding out the first few letters. Play the matching game a few times!

PLANT VOCABULARY

Carbon Dioxide:

A chemical in the air that Plants breathe in, and Humans breathe out.

Oxygen:

A chemical in the air that humans breathe in and plants breathe out.

PhotoSynthesis
How plants live, eat, and make energy.

Chlorophyll:
The chemical that makes plants GREEN.

Life Cycle:
A plants life from planting, to growing, to breaking down and becoming soil.

PLANT VOCABULARY

Practice these daily.
You can learn none of these; one of these; or all of these.

If you learn 1 or 2, this Unit could take 1 week.

If you learn 3-5, this Unit could take up to a month.

Remember: Go at your child's pace.

PLANT FACTS

1. Plants eat using Photosynthesis. They convert light energy from the sun, into energy they can use for food.

2. Most plants need plenty of sunlight, good soil, and water.

3. Plants are living things.

4. There are more than 250,000 types of Flowers.

5. There are over 20,000 species of edible plants in the world. We generally only eat **20.**

PLANT LINES

Stay on the line, try to get the bug to the other side!

PLANT TRACING

PLANT TRACING

PLANT TRACING

Flower

PARTS OF A FLOWER

- Stem
- Leaf
- Roots
- Petals

Cut and paste them in the boxes above.

PARTS OF A FLOWER

PARTS OF A TREE

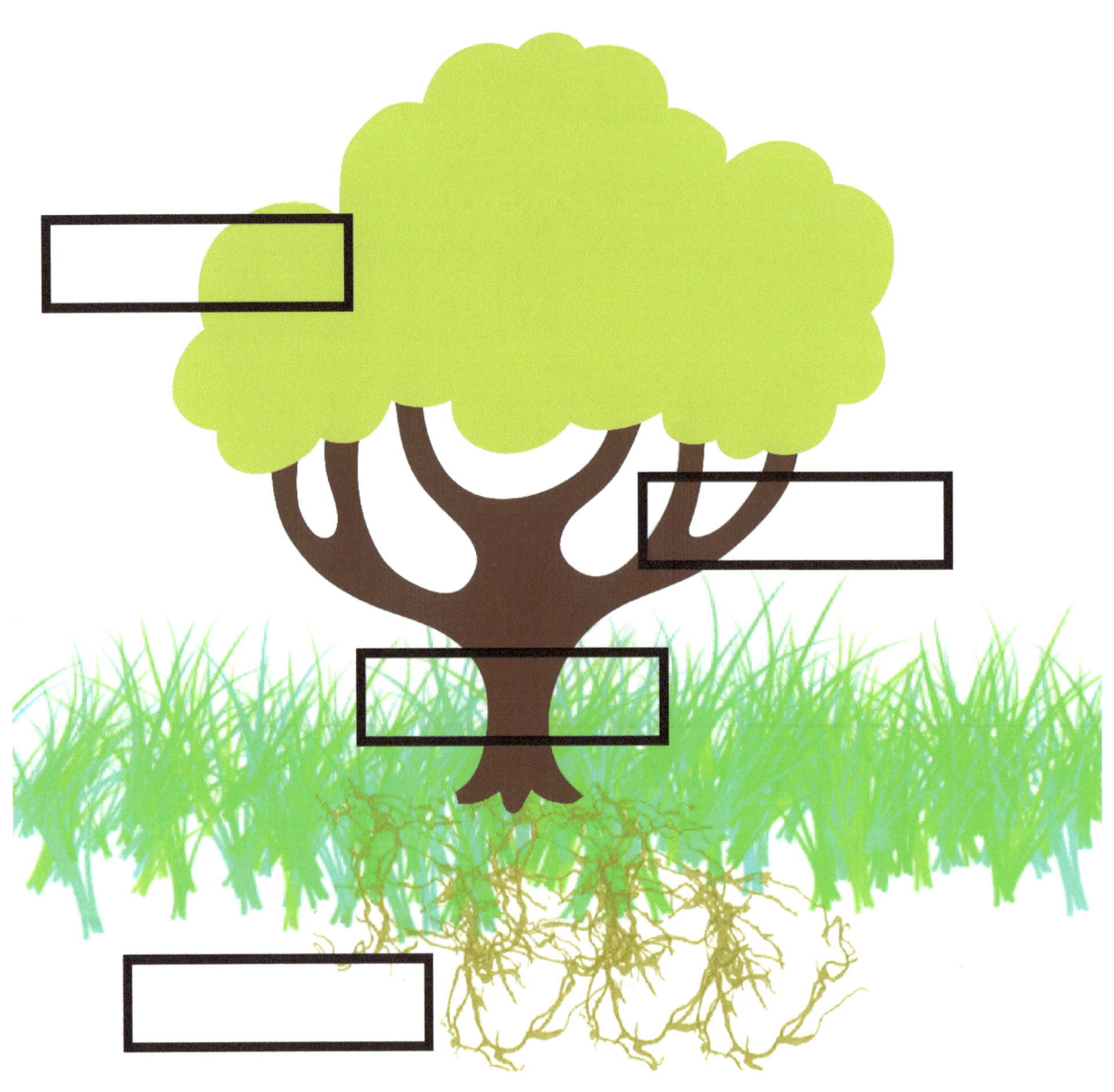

PARTS OF A TREE

Trunk

Leaf

Roots

Branches

Cut and paste them in the boxes above.

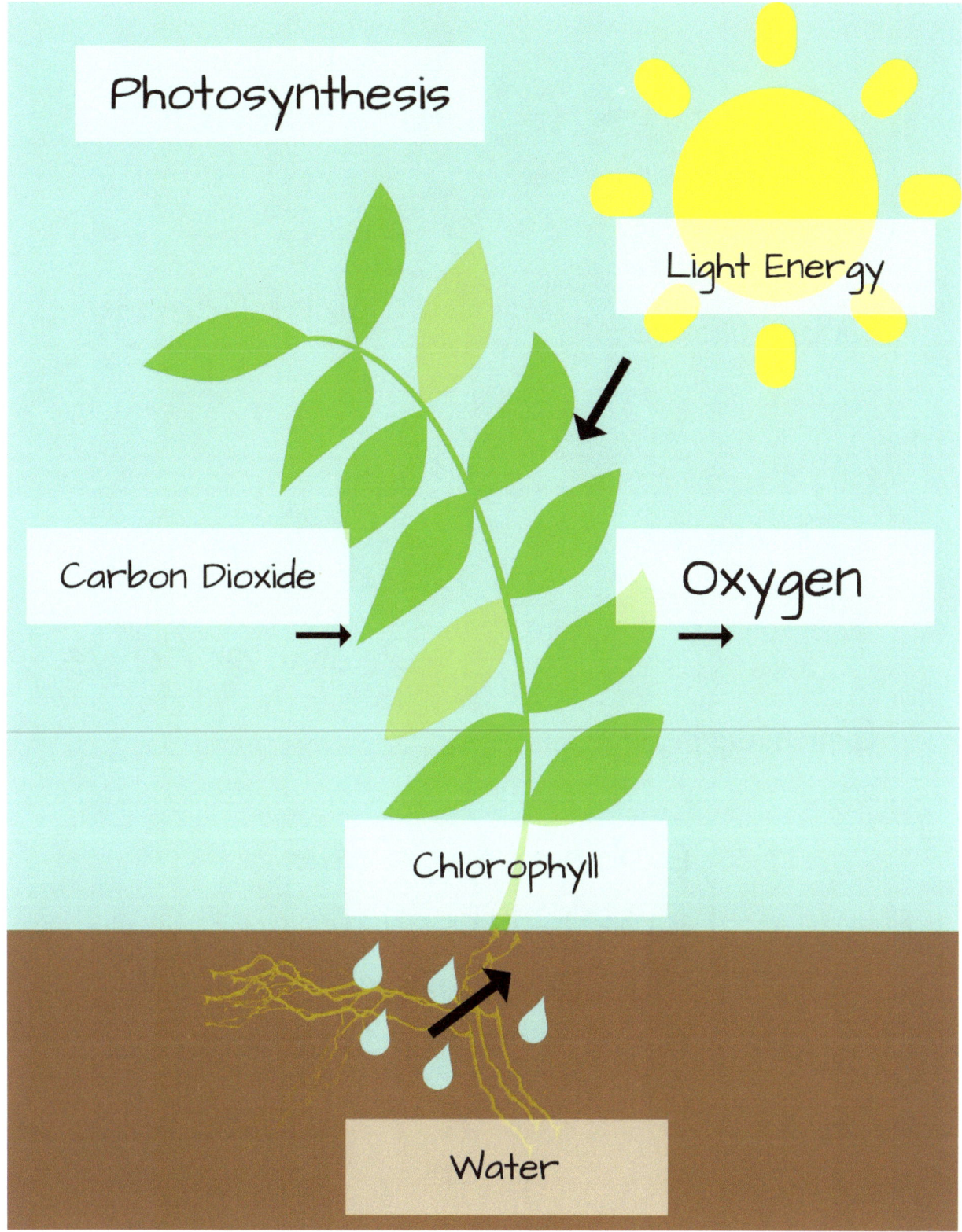

PhotoSynthesis

Water

Carbon Dioxide

Light Energy

Oxygen

Cut and paste, Draw arrows.

Chlorophyll

Explain to your child:
1. Plants eat sunlight, drink water, and breathe in Carbon Dioxide
2. They breathe out Oxygen, and make sugar.
3. They have Chlorophyll which makes them green!

TREE LIFECYCLE

TREE LIFECYCLE

Instructions:

Cut and paste
the steps of the Lifecycle
in order
On the page that just says
Tree Lifecycle

The steps are, just llike flowers:
1. Good Soil
2. Plant Seeds
3. Buds form
4. Grows little plant
5. Grows into a tree or flower
6. One day decomposes and starts all over.

PLANT MATCHING

 Leaf

 Flower

 Tree

 Sun

PLANT MATH

Place the between 5 and 20 objects on the picture based on the number shown.

5

PLANT MATH

10

PLANT MATH

PLANT MATH

20

PLANT MATH

- Place an object on each leaf. Try to keep them in the lines!

NATURE HUNT- 5 SENSES

Go on a Scavenger Hunt. Find something that looks like a plant, tastes like its from a plant, sounds like nature, feels like a plant, and smells like a plant. X out when you are done.

PLANT BINGO

Roll a die. Count the dots, and put a marker on the corresponding picture. Play til someone wins!
You may need to roll the die twice, if so it is a good math lesson!
Addition.

PLANT TIC TAC TOE

PLANT TIC TAC TOE

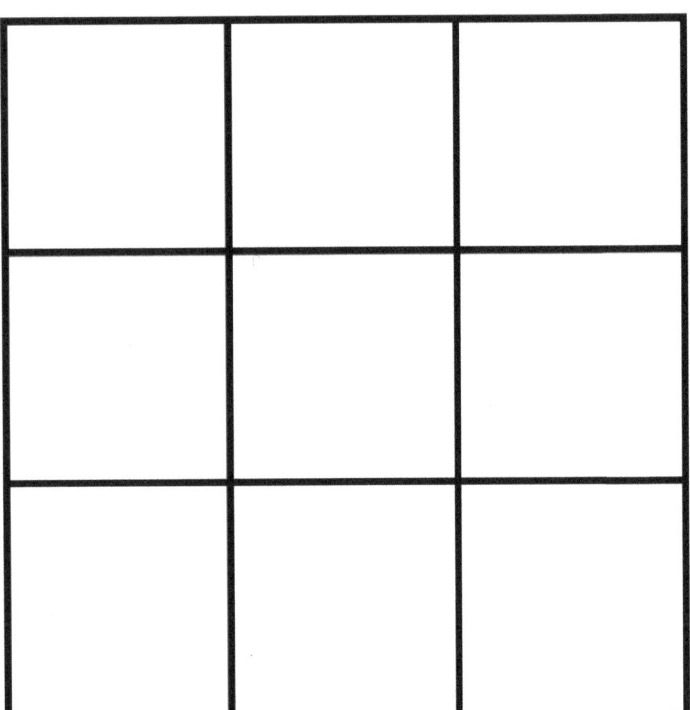

These are your x's and o's. Cut them out, and you can have up to 3 players.

PLANT RECIPE

From your garden, the store, or Farmer's Market

<u>Homemade Tomato Sauce</u>

2 tbsp each of

Oregano

Thyme

Marjoram

Rosemary

Chives

Parsley

8-12 tomatoes

1/2 cup water

1 tbsp sugar to cut the acidity

PLANT RECIPE

Directions:

1. Chop the herbs finely.
2. Dice the tomatoes.
3. Add all ingredients to a pot, simmer for an hour.
4. Puree
5. Tomato sauce!

Follow us!
Raisinghumanbeans.com
Facebook.com/raisingbeans
Instagram.com/raisinghumanbeans
Pinterest.com/raisinghumanbeans

© Raising Human Beans
2018

www.ingramcontent.com/pod-product-compliance
Lightning Source LLC
Chambersburg PA
CBHW042323250526
R18347300001B/R183473PG45473CBX00020B/15